# Inventions 100 Past Present:

## *The TNT Group*

# Prelude

Inventions have been and continue to be beneficial for mankind in many ways:

- **Improving quality of life:** Inventions have made it possible for us to live longer, healthier, and more comfortable lives. They have provided us with better healthcare, transportation, communication, and entertainment.

- **Enhancing productivity:** Inventions have helped us to be more productive and efficient in our work. Machines, tools, and equipment have made it easier and faster to carry out tasks that would have taken us a lot of time and effort to complete manually.

- **Driving economic growth:** Inventions have played a crucial role in economic growth by creating new industries, generating employment opportunities, and increasing the standard of living. Many of the most successful companies in the world today were built on the back of innovative ideas.

- **Advancing science and knowledge:** Inventions have led to new discoveries and breakthroughs in various fields such as medicine, engineering, physics, and computer science. They have expanded our understanding of the world and paved the way for further advancements.

Overall, inventions have been and continue to be good for mankind as they have improved our lives, enhanced our productivity, driven economic growth, and advanced science and knowledge.

# All Images Included

Content License[1] (pixabay.com)

Free to use under the Content License (All Images rendered from Pixabay.com)

No attribution required

© 2023, March,TheTNTGroup

---

## Light bulbs

The modern electric light bulb was not invented by a single person, but rather by several inventors who made important contributions to its development.

Thomas Edison is often credited with inventing the first commercially practical incandescent light bulb. In 1879, he filed a patent for an improved version of the incandescent lamp that used a carbon filament in a vacuum-sealed bulb. This design made the bulb more durable and longer-lasting than previous versions.

However, Edison was not the first to invent the incandescent light bulb. Other inventors, including Joseph Swan in the UK and Hiram Maxim in the US, had also developed early versions of the incandescent lamp. Swan received a patent for his design in 1878, a year before Edison's patent.

So while Edison is often associated with the invention of the light bulb, it is important to recognize the contributions of other inventors who helped to develop this important technology.

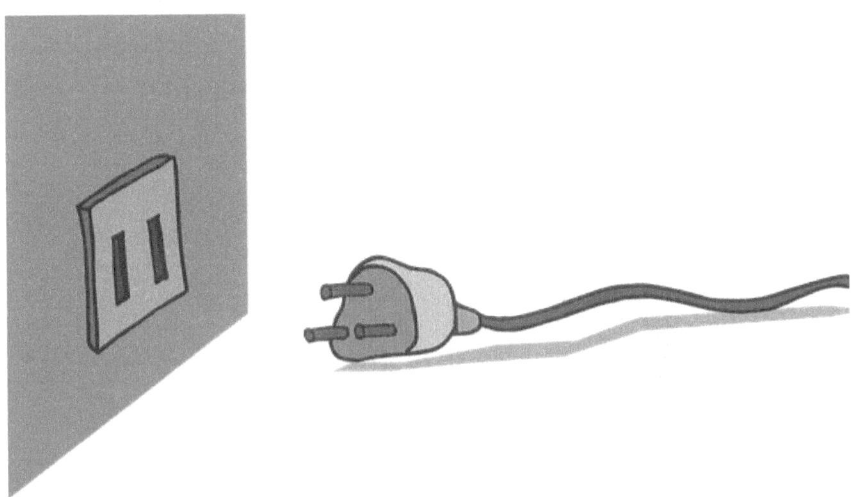

## Outlets

The modern electric socket, also known as an electrical outlet or power outlet, was not invented by a single person. Rather, it was the result of many years of technological development and refinement by numerous inventors and engineers.

One of the earliest versions of the electrical outlet was developed by Thomas Edison in the late 1800s, as part of his efforts to develop a safe and reliable electrical distribution system. However, Edison's design was not the first electrical outlet, and it was not widely adopted.

Over the years, many other inventors and engineers contributed to the development of the modern electrical outlet. In 1904, Harvey Hubbell patented the first detachable electrical plug, which made it easier to connect and disconnect devices from electrical outlets. In 1915, General Electric introduced the first two-prong electrical outlet, which became the standard design for many years.

Since then, many improvements have been made to the electrical outlet, including the addition of a third grounding prong for improved safety and the development of smart outlets that can be controlled by smartphones and other devices.

So, while there isn't a single inventor who can be credited with inventing the modern electrical outlet, it was the result of many years of technological development and innovation by numerous inventors and engineers.

**Popcorn**

The exact origin of popcorn is not known, but it is believed that popcorn has been around for thousands of years. Archaeological evidence suggests that popcorn was cultivated by ancient civilizations in the Americas, including the Aztecs and the Incas.

However, popcorn, as we know it today, was first popularized in the late 19th century by Charles Cretors, a popcorn machine inventor and entrepreneur from Chicago. Cretors developed a steam-powered popcorn machine in 1885 that could pop large quantities of popcorn quickly and efficiently. He also developed a special caramel coating that made popcorn more flavorful and appealing to consumers.

Thanks to Cretors' innovations, popcorn quickly became a popular snack food in America and worldwide. Today, popcorn is enjoyed by millions of people and is available in a wide variety of flavors and forms.

### Ice cream

The origins of ice cream are not entirely clear, as there is evidence that various forms of frozen desserts were consumed in ancient China, Persia, and Rome. However, it is generally believed that the modern version of ice cream as we know it today was developed in Italy during the Renaissance.

One of the early pioneers of ice cream was Catherine de' Medici, a member of the Italian noble family who became the Queen of France in the 16th century. She is credited with introducing ice cream to the French court, where it became a popular delicacy.

The first recorded mention of ice cream in English was in the 18th century, when it was served at a banquet in honor of the new governor of Maryland. Ice cream was not widely available to the general public until the mid-19th century when technological advances made it easier and cheaper to produce.

Today, ice cream is one of the most popular desserts in the world, enjoyed in a wide variety of flavors and forms. It is typically made from milk, cream, sugar, and flavorings, and is frozen to create a smooth, creamy texture.

## Skiing

The origins of skiing are not entirely clear, as it is believed to have been developed independently in various parts of the world over thousands of years. However, it is generally accepted that skiing has its roots in Scandinavia, where people used skis for transportation, hunting, and warfare.

The oldest known skis, dating back to around 6000 BCE, were discovered in northern Russia. These skis were made from wood and were used by early hunters and nomads to travel across snowy terrain. Similar ski-like artifacts have been found in other parts of northern Europe, including Norway and Sweden.

Skiing as a sport began to develop in the late 19th century in Norway, where organized ski races and jumping competitions were held. Skiing quickly spread to other parts of Europe and North America, and today it is enjoyed by millions of people around the world as both a recreational activity and a competitive sport.

## Snowboarding

Snowboarding was invented in the 1960s and 1970s by several individuals who experimented with different designs and techniques for riding on snow. However, the modern form of snowboarding, as we know it today, is largely credited to Sherman Poppen, who created the first snowboard prototype in 1965.

Poppen's invention was called the "Snurfer," which was a combination of "snow" and "surfer." The Snurfer consisted of two skis bolted together and a rope attached to the front for balance and control. The Snurfer gained popularity in the 1960s and led to the development of modern snowboards in the 1970s.

Over time, many other individuals contributed to the development of snowboarding as a sport, including Tom Sims, Jake Burton Carpenter, and Dimitrije Milovich. Today, snowboarding is a widely popular winter sport enjoyed by millions of people around the world.

## Chocolate

The history of chocolate dates back to the ancient Mesoamerican civilizations, including the Aztecs and Mayans, who consumed chocolate as a bitter beverage made from ground cacao beans. However, it is difficult to attribute the invention of chocolate to a single individual, as it evolved over time through a series of cultural and culinary influences.

The first known use of cacao beans dates back to 1900 BCE, but it wasn't until around 600 CE that the Mayans began cultivating and processing the beans into a paste that could be mixed with water and spices to make a frothy beverage. The Aztecs later adopted this practice and added their own twist by sweetening the drink with honey.

The Europeans discovered chocolate when the Spanish conquistadors arrived in the Americas in the 16th century. They brought back cacao beans to Europe and soon developed a taste for the bitter beverage, which they sweetened with sugar.

Over time, chocolate evolved into the forms we know today, such as chocolate bars, truffles, and other confections. But its origins can be traced back to the ancient Mesoamerican cultures who first cultivated and consumed the cacao bean.

## Coffee

The exact origin of coffee is not known, but it is believed to have originated in Ethiopia, where coffee trees still grow wild today. According to legend, a goat herder named Kaldi in Ethiopia noticed that his goats became energetic after eating the berries from a certain tree. Curious, he tried the berries himself and experienced a similar effect.

The coffee plant was then cultivated and spread to other parts of the world, including the Arabian Peninsula, where it was first roasted and brewed as a drink. It eventually became popular throughout the Islamic world and was brought to Europe by merchants in the 16th century.

It is difficult to attribute the invention of coffee to a single person as its origins are shrouded in legend and the drink has evolved over centuries. However, we do know that coffee has a long and rich history that spans many cultures and countries.

**Bungee Jumping**

Bungy jumping was first commercialized by AJ Hackett and Henry van Asch of New Zealand in the 1980s. The activity involves jumping from a high platform or bridge while attached to an elastic cord. Today, bungy jumping is a popular adventure activity around the world.

The modern version of bungee jumping, as a recreational activity, was invented by a group of students from the University of Oxford in England in 1979. The group, led by a man named David Kirke, had heard about the ritualistic land diving performed by the people of Pentecost Island in Vanuatu, where men jump from tall wooden platforms with vines tied to their ankles as a rite of passage.

The Oxford students were inspired by this and decided to try jumping from a bridge with elastic cords tied to their ankles. They carried out their first jump from the 250-foot Clifton Suspension Bridge in Bristol, England, and the activity quickly gained popularity as an extreme sport.

Although the Oxford students are credited with inventing modern bungee jumping, there have been records of similar activities being carried out in different parts of the world for centuries, such as the aforementioned land diving in Vanuatu and similar rituals in other Pacific Island cultures.

**Yoga**

The practice of yoga has evolved over thousands of years and its exact origin is not attributed to any single individual. It is believed to have developed in ancient India, with the earliest known written record of yoga techniques dating back to around 3,000 BCE in the Indus Valley civilization.

The ancient Indian sage Patanjali is credited with compiling the foundational text of yoga, the Yoga Sutras, around the 2nd century BCE. This text outlines the philosophy and practices of yoga, including meditation, breathing techniques, and ethical principles.

Over time, yoga has continued to evolve and adapt, with different schools and styles emerging. Today, yoga is practiced all around the world and has become a popular form of exercise, meditation, and spiritual practice for millions of people.

## Cars

The invention of the automobile, or the car, is generally credited to Karl Benz, a German engineer who patented the first gasoline-powered automobile in 1886. However, the development of the automobile was the result of the work of many inventors and engineers over several decades.

In the early 19th century, several inventors, including Nicolas-Joseph Cugnot and Richard Trevithick, developed steam-powered vehicles, which were the precursors to the modern automobile. In the late 19th century, several other inventors, including Gottlieb Daimler and Wilhelm Maybach, also made significant contributions to the development of the automobile.

However, it was Karl Benz who is credited with inventing the modern automobile with his gasoline-powered vehicle, the Benz Patent-Motorwagen, which was the first vehicle to be designed specifically for use on public roads. The Benz Patent-Motorwagen had many features that are still found in modern cars, including a gasoline-powered internal combustion engine, a carburetor, and a differential.

### Electric cars

Electric cars have a long history, and there were many inventors and scientists who contributed to their development over time. However, the first practical electric car was invented in 1884 by Thomas Parker, a British inventor and engineer.

Parker was an early advocate of electric vehicles, and he built his first electric car in 1884. The car was powered by a rechargeable battery, and it was capable of reaching speeds of up to 18 miles per hour.

Over the next few decades, electric cars became more popular, particularly in urban areas where their quiet operation and lack of pollution made them attractive to city dwellers. However, the invention of the internal combustion engine in the late 19th century led to the widespread adoption of gasoline-powered cars, and electric cars fell out of favor.

In recent years, electric cars have made a comeback, thanks in part to advances in battery technology and a growing concern about the environmental impact of fossil fuels. Today, many companies, including Tesla, Nissan, and General Motors, produce electric cars for the mass market.

### Bitcoin

The identity of the person or group of people who invented Bitcoin is still unknown, and they used the pseudonym "Satoshi Nakamoto." In 2008, Satoshi Nakamoto released a white paper titled "Bitcoin: A Peer-to-Peer Electronic Cash System," which described a decentralized digital currency that uses cryptography to secure transactions and control the creation of new units. The first Bitcoin software was released in 2009, and Nakamoto continued to contribute to the development of the cryptocurrency until 2010, when they disappeared from public view and handed over control of the project to others. Despite several attempts to uncover the true identity of Satoshi Nakamoto, their real name and location remain a mystery.

### Stock Market

The stock market as we know it today did not have a single inventor. Instead, it evolved over time through various innovations and historical events.

One of the earliest examples of a stock market can be traced back to 17th-century Amsterdam, where the Dutch East India Company issued shares of stock to the public to finance its operations. This allowed individuals to invest in the company's profits and helped spread the financial risk among a larger group of investors.

In the United States, the New York Stock Exchange (NYSE) was founded in 1817 and became a central hub for buying and selling stocks. However, there were already several other stock exchanges operating in the U.S. at the time, such as the Philadelphia Stock Exchange and the Boston Stock Exchange.

The modern stock market has continued to evolve and expand over time, with the advent of electronic trading systems, new financial instruments, and changes in regulations and governance. Therefore, it is difficult to attribute the invention of the stock market to a single individual or group.

### Eyeglasses

The invention of eyeglasses is generally attributed to an Italian named Salvino D'Armate, who is believed to have invented them around the year 1284. However, there is some debate among historians as to whether D'Armate was actually the first person to invent eyeglasses, as there is evidence that similar devices may have existed in other parts of the world at earlier times.

Regardless of who was the first to invent eyeglasses, it is clear that they have been an incredibly important invention that has had a profound impact on society. Eyeglasses have enabled millions of people around the world to see more clearly, and they have played a critical role in many fields, including science, medicine, and engineering. Today, eyeglasses are available in a wide variety of styles and designs, and they continue to be an essential tool for people of all ages and backgrounds.

**Contact Lens**

The concept of contact lenses can be traced back to the late 19th century, but the first person to create a wearable contact lens was a German glassblower and optician named Adolf Fick. In 1887, Fick developed a contact lens made of blown glass that he successfully fitted onto his own eye. However, Fick's design was not practical for widespread use, as the glass lenses were uncomfortable to wear and could only be worn for a few hours at a time.

The first contact lenses made of plastic were invented by Czech chemist Otto Wichterle in the early 1950s. Wichterle developed a technique for creating soft contact lenses using a hydrogel material called HEMA (2-hydroxyethyl methacrylate), which was more comfortable and flexible than the hard glass lenses developed by Fick.

Since then, contact lens technology has continued to evolve, with advances in materials, design, and manufacturing processes leading to a wide range of options for people with vision problems. Today, contact lenses are a popular alternative to eyeglasses for people who prefer the convenience and aesthetic benefits of wearing contact lenses.

## Dice

The exact origins of dice are unclear, as they have been used for thousands of years in various cultures around the world. The earliest known dice were excavated from an archaeological site in southeastern Iran, dating back to around 2800 BC.

It is believed that dice were originally used for divination and fortune-telling, rather than for gaming. They were also used in ancient Egypt and Greece for various purposes, such as determining outcomes in legal disputes and selecting officials by lot.

It is difficult to attribute the invention of dice to a specific individual or culture, as the use of dice has been widespread throughout history. However, it is clear that dice have played an important role in human history and have been used for a variety of purposes for thousands of years.

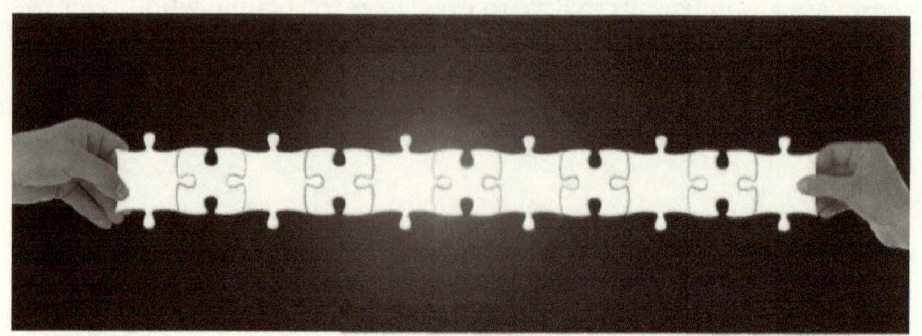

### Puzzles

The origin of puzzles is not clear, as they have been used for thousands of years in various cultures around the world. Ancient Chinese and Egyptian cultures used puzzles for entertainment and educational purposes. For example, Chinese Tangram puzzles date back to the Song Dynasty (960-1279 AD), while Egyptian hieroglyphics depict various types of puzzles, such as mazes and riddles.

In Europe, puzzles became popular in the 18th and 19th centuries. The first jigsaw puzzle was created by John Spilsbury, a British mapmaker, in the 1760s. He created a puzzle by cutting a map into pieces to teach geography to his students. Later, puzzles became more sophisticated and intricate, with the invention of the puzzle box, the Rubik's Cube, and other mechanical and electronic puzzles.

Overall, the exact inventor of puzzles is unknown, as puzzles have been used for thousands of years and have evolved over time. Nonetheless, puzzles have been enjoyed by people all over the world for their entertainment and educational value.

## Jeans

Jeans are a type of pants made from denim fabric that was first developed in the 19th century. The modern design of jeans is attributed to Levi Strauss, a German-born American businessman who founded Levi Strauss & Co. in San Francisco in 1853.

Levi Strauss originally sold dry goods, including canvas for wagon covers and tents, to prospectors during the California Gold Rush. In the 1870s, he began to make pants out of denim fabric, which was a durable and practical material for miners and other laborers.

Levi Strauss partnered with a tailor named Jacob Davis, who had developed a method of reinforcing the pockets and stress points of pants with metal rivets. Together, they obtained a patent for the design in 1873 and began producing "waist overalls," which later became known as "jeans."

Jeans grew in popularity throughout the 20th century, especially among young people, and became an iconic symbol of American culture. Today, jeans are worn around the world and come in a wide range of styles, colors, and designs.

### High Heels

The invention of high heels is attributed to various civilizations throughout history, including ancient Egypt, Greece, and Rome. However, it is believed that the modern high heel, as we know it today, was popularized in the 17th century by French King Louis XIV.

Louis XIV was known for his love of fashion and luxury, and he often wore high-heeled shoes with red soles to make himself appear taller and more imposing. The style quickly caught on among the French aristocracy and became a symbol of wealth and status.

While Louis XIV did not technically invent high heels, he certainly played a significant role in popularizing the style and making it a fashion statement.

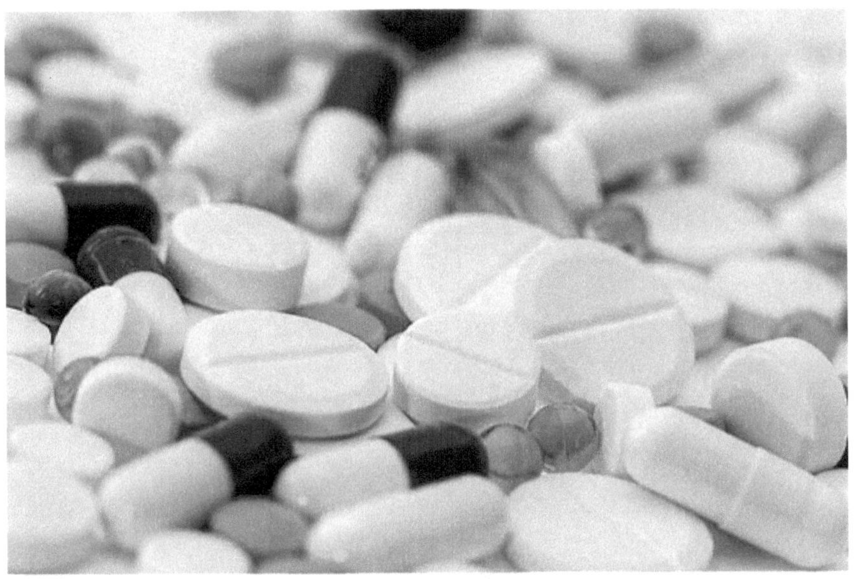

### Painkillers

The use of painkillers dates back thousands of years, and it is difficult to attribute their invention to a single individual. In ancient times, people used various natural substances, such as opium and willow bark, to relieve pain.

However, the modern era of painkillers began in the 19th century with the discovery of morphine, a powerful pain-relieving drug derived from opium. Morphine was first isolated by the German pharmacist Friedrich Sertürner in 1804.

Aspirin, one of the most widely used painkillers today, was invented in 1897 by the German chemist Felix Hoffman, who was working for the pharmaceutical company Bayer. Hoffman synthesized a new version of salicylic acid, which had been used for centuries as a pain-relieving substance found in willow bark, and named it "aspirin."

Since then, many other painkillers have been developed, including acetaminophen, ibuprofen, and naproxen, among others.

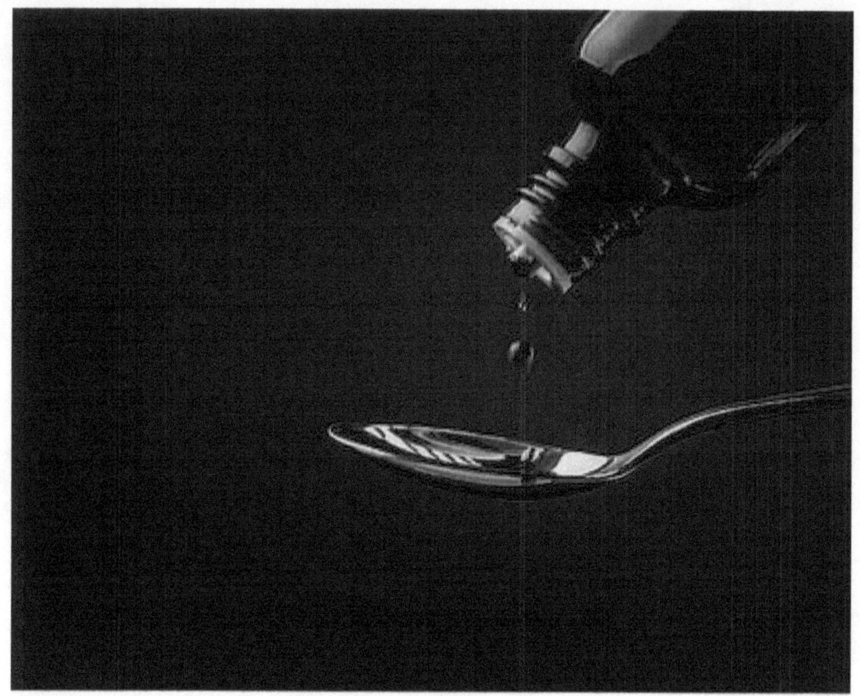

## Cough Syrup

The use of cough syrup dates back to ancient times, when people would use various natural remedies to alleviate coughs and other respiratory symptoms. However, it is difficult to pinpoint a single inventor of cough syrup, as the use of such remedies has been documented in many cultures throughout history.

In modern times, cough syrups have been developed and manufactured by pharmaceutical companies. Some of the early cough syrups were patented in the 19th century, and many of these early formulations contained ingredients such as morphine or codeine, which are now highly regulated due to their addictive properties.

Today, cough syrups are typically formulated with a combination of active ingredients, such as dextromethorphan (a cough suppressant) and guaifenesin (an expectorant), along with various other ingredients to provide flavor and texture. The exact composition of cough syrup can vary widely depending on the manufacturer and the intended use.

## The Piano

The modern piano as we know it today was invented by an Italian musical instrument maker named Bartolomeo Cristofori (1655-1731) in the early 18th century. Cristofori is credited with inventing the first hammer mechanism that allowed for dynamic control of the instrument's volume. He called his invention the "gravicembalo col piano e forte," which means "harpsichord with soft and loud" in Italian.

The first piano built by Cristofori is believed to have been made around the year 1700. Over the next few decades, he continued to refine and develop his invention, producing pianos with improved sound quality, range, and playability. Despite the popularity of the piano today, it took several decades for it to gain widespread acceptance as a musical instrument.

### The Guitar

The modern guitar as we know it today was developed in the 19th century, but its origins can be traced back to ancient civilizations. It is believed that the earliest form of the guitar was a stringed instrument called the tanbur, which was used in Persia around 3000 years ago. The Guitar - Al-Farabi: Al-Farabi, a Persian philosopher and musician, is credited with inventing the guitar in the 10th century.

The guitar as we know it today, with six strings and a fretted fingerboard, is generally credited to a Spanish luthier named Antonio de Torres Jurado, who lived from 1817 to 1892. He is considered the father of the modern classical guitar and is responsible for many of the design innovations that have made the guitar what it is today.

However, it's important to note that many different cultures around the world have contributed to the development of stringed instruments that resemble the modern guitar, and it's likely that the instrument evolved over time through the contributions of many different people and cultures.

### License Plates

The concept of license plates, which are used to identify and register vehicles, was first introduced in France in 1893. The idea was proposed by a French politician named Paul Faure, who suggested that all motor vehicles should be required to display a unique identification number.

The first license plate was a simple metal plaque that was affixed to the back of the vehicle, and it was issued by the police department in Paris. Other countries soon followed suit, and by the early 20th century, license plates had become standard practice in most parts of the world.

The United States began using license plates in 1901, starting with the state of New York. The first plates were made of iron and were issued to only a handful of vehicle owners, as automobiles were still a relatively new and expensive technology at the time. Over time, the use of license plates became more widespread and uniform across the United States and other countries.

### Jump Rope

The exact origins of jump rope are not entirely clear, as various forms of rope jumping have been practiced for centuries by different cultures around the world. However, it is believed that the modern form of jump rope, using a long, lightweight rope and performed as a game or sport, was first popularized in the United States in the 19th century.

Some sources credit the British as inventors of jump rope, while others suggest that it was African-American slaves who first used ropes to exercise and entertain themselves during their captivity. In any case, jump rope quickly became a popular activity among children, especially girls, in the late 1800s and early 1900s.

The first patented jump rope was created by a man named W.P. Emerson in 1866, but the basic design of the rope and the game has remained largely the same over the years, with variations in the types of ropes and techniques used.

## Umbrellas

The umbrella is an ancient invention, with evidence of its use dating back to ancient Egypt, Greece, and Rome. However, the modern collapsible umbrella that we use today was invented in the 16th century in Europe.

The first recorded use of an umbrella in Europe was by a woman named Mariotto degli Strozzi, who used one in 16th-century Italy. The English claim to have invented the modern umbrella, as the first patent for a folding umbrella was granted to a London merchant named Jonas Hanway in 1786. However, other inventors in France and Germany were also working on similar designs around the same time.

Regardless of who can claim credit for inventing the umbrella, it has become an essential item for protecting us from the rain and sun.

### Rollercoasters

The modern roller coaster as we know it was developed in the United States in the late 19th century. The first roller coaster was designed and built by LaMarcus Adna Thompson in 1884 and it was called the "Switchback Railway."

The Switchback Railway was a simple coaster that traveled at speeds of only six miles per hour and did not have any of the loops or inversions that are common in modern coasters. However, it was an immediate success and inspired other inventors to develop more thrilling and complex coasters in the years that followed.

Today, roller coasters are popular attractions in amusement parks around the world, and they continue to evolve and push the boundaries of what is possible in terms of speed, height, and excitement.

## The Laundry Machine

The invention of the laundry machine is credited to a number of people who developed and improved washing machines throughout history. The first washing machines were hand-powered and used for washing clothes by hand.

In the 18th century, the first mechanical washing machine was invented by Nathaniel Briggs, an American. However, it was not widely adopted.

The first commercially successful washing machine was patented by Hamilton Smith in 1858. It was a rotary washing machine that used a hand-operated crank to agitate the clothes in a drum filled with water and soap.

In the 20th century, many improvements were made to washing machines, including the addition of electric motors, automatic timers, and agitators. One of the most significant advances was the introduction of the automatic washing machine, which was first developed in the 1930s by companies such as Bendix and General Electric.

Today, washing machines are an essential appliance in most households, and they continue to be improved upon with new technologies such as energy-efficient features, smart controls, and even robotic washing machines.

## Bibs

The use of bibs for babies dates back centuries, as mothers have always found ways to protect their infants' clothing during feeding. However, the specific invention of the modern baby bib is not attributed to any one person.

In the early 20th century, the first commercially available baby bibs were made of rubber and were intended to be wiped clean. Later, cloth bibs with snap closures or tie strings became popular. Today, there are many different types of baby bibs available, including disposable bibs, silicone bibs, and bibs with food-catching pockets.

While the exact inventor of the baby bib is unknown, it is clear that the idea of protecting a baby's clothing during feeding has been around for centuries and has evolved over time to meet the changing needs of parents and infants.

### Baby Car Seats

The invention of the first baby car seat is attributed to Jean Ames, a mechanical engineer, who designed a rear-facing child restraint in 1933. However, this early design was not widely adopted, and it wasn't until the 1960s that car seats for children became more widely used and regulated. In 1962, an American named Leonard Rivkin designed the first commercial baby car seat, which was called the "Tot-Guard." The design was a simple metal frame with a canvas cover and was designed to be strapped to the car's seat. Since then, baby car seats have undergone significant improvements and regulatory changes to improve their safety and effectiveness.

## Gold

Gold has been known and valued by humans for thousands of years, and its discovery cannot be attributed to a single person or culture. Gold was likely discovered through observations of its bright, yellow color and its relative rarity in the earth's crust.

Archaeological evidence suggests that ancient civilizations in Egypt, India, and Mesopotamia were mining and using gold as early as 2600 BCE. The ancient Greeks and Romans also valued gold and used it for decorative purposes and as currency.

The discovery of gold in the Americas in the 15th and 16th centuries played a significant role in the colonization and exploration of the New World. Spanish conquistadors discovered large amounts of gold in Central and South America, which fueled their expeditions and helped to establish their empires.

Today, gold remains a valuable and sought-after metal, used in a wide range of applications, from jewelry and art to electronics and medicine.

## Diamonds

The discovery of diamonds is not attributed to a single individual, as diamonds have been known and traded for thousands of years. However, the first recorded discovery of diamonds was in India, where they were mined and traded as early as the 4th century BCE. Diamonds were highly valued for their beauty and durability and were used in jewelry, as well as for cutting and polishing other gems.

In the 18th and 19th centuries, significant diamond deposits were discovered in Brazil, Russia and South Africa, leading to a surge in the supply of diamonds and making them more widely available. Today, diamonds are still highly prized and are mined in various countries around the world.

## The Motorcycle

The invention of the motorcycle is attributed to several individuals who developed early versions of motorized bicycles in the late 19th century. However, the first commercially successful motorcycle was produced by the German company Hildebrand & Wolfmüller in 1894.

The Hildebrand & Wolfmüller motorcycle was powered by a two-cylinder, four-stroke engine and had a top speed of 28 miles per hour (45 kilometers per hour). It was the first motorcycle to be produced in large numbers and sold around the world.

Other early motorcycle pioneers include Gottlieb Daimler and Wilhelm Maybach, who developed a gasoline-powered engine in 1885 that was used in a motorized bicycle; Sylvester Roper, an American inventor who built steam-powered motorcycles in the 1860s and 1870s; and Edward Butler, an English inventor who produced a gasoline-powered tricycle in 1884.

Over the years, motorcycles have continued to evolve and improve, with new technologies, designs, and features. Today, motorcycles are popular around the world for transportation, recreation, and sport.

## Trains

Trains as a form of transportation have a complex history with many inventors and engineers contributing to its development over time.

The earliest known form of a train was developed in ancient Greece, where a toy called a "peripatoi" was created, which consisted of a small cart that was pulled along a track by a cord. However, it wasn't until the Industrial Revolution in the 18th and 19th centuries that trains began to be developed as a practical means of transportation.

One of the key figures in the development of the modern train was George Stephenson, an English engineer who is often referred to as the "Father of Railways." In 1814, Stephenson built the world's first steam locomotive, called the Blücher, which was used to haul coal from a mine in England. Stephenson continued to refine his designs, and in 1825, he built the world's first public railway, the Stockton and Darlington Railway.

Another important figure in the development of trains was Robert Stephenson, George Stephenson's son, who built the Rocket, a steam locomotive that was used on the Liverpool and Manchester Railway. The Rocket set the standard for steam locomotives and was the first to use a multi-tube boiler, which greatly increased its efficiency.

Throughout the 19th and 20th centuries, trains continued to evolve, with new technologies such as diesel and electric power being developed. Today, trains remain an important form of transportation around the world, and their invention and development was the result of the work of many inventors and engineers over several centuries.

### Bullet Trains

The bullet train, also known as the Shinkansen, was invented by a team of engineers led by Hideo Shima in Japan in the early 1960s. The first bullet train line, between Tokyo and Osaka, opened in 1964, just in time for the Tokyo Olympics. Since then, the bullet train has become an iconic symbol of Japan's technological innovation and efficiency, and has been widely emulated in other countries around the world.

## Tractors

The invention of the tractor was a gradual process involving many inventors and innovators. The first steam-powered tractor was invented by Richard Trevithick in 1812, and the first gasoline-powered tractor was invented by John Froelich in 1892. However, it was the American inventor, Charles W. Hart and Charles H. Parr, who invented the first gasoline-powered tractor with an internal combustion engine in 1902. Their tractor was called the "Hart-Parr Gasoline Engine Tractor" and was manufactured by the Hart-Parr Company. Over time, many other inventors and companies contributed to the development and improvement of tractors, making them more powerful, efficient, and versatile. Today, tractors are widely used in agriculture, construction, and transportation industries.

## The Airplane

The first international flight is generally considered to have taken place on August 17, 1919, when a flight operated by British airline Aircraft Transport and Travel (AT&T) flew from Hounslow Heath Aerodrome near London, England to Le Bourget Airport near Paris, France. The flight carried one passenger, a journalist named H.E. Ponsford, and took approximately 2 hours and 30 minutes. This historic flight marked the beginning of international commercial air travel, which has since become a vital aspect of global transportation and commerce.

The Wright brothers, Orville and Wilbur Wright, are credited with inventing and building the world's first successful airplane. They achieved this feat on December 17, 1903, when they flew their Wright Flyer aircraft for 12 seconds over a distance of 120 feet at Kitty Hawk, North Carolina, USA. This historic event marked the beginning of the aviation era, and the Wright brothers' contributions to the field of aeronautics are widely recognized and celebrated.

The Tiltrotor Aircraft: Brazilian aeronautical engineer, Frank Robinson, designed and patented the first tiltrotor aircraft in 1960.

## Wind Turbines

The Urban Wind Turbine: In 2004, Brazilian inventor and entrepreneur, Júlio César Rodrigues, created the first urban wind turbine. The device can generate electricity from the wind in urban environments, making it a promising source of renewable energy for cities.

The invention of wind turbines is attributed to several individuals and civilizations over time.

One of the earliest known uses of wind power was by the ancient Persians who used windmills to grind grain in the 7th century AD.

In the late 19th century, American engineer Charles F. Brush designed a wind turbine to generate electricity, which he installed at his home in Cleveland, Ohio in 1888.

However, it was Danish scientist and inventor Poul la Cour who is considered one of the pioneers of modern wind turbine technology. In the early 20th century, la Cour developed wind turbines with rotor blades that faced into the wind, a design that is still used today.

Today, wind turbines are used around the world to generate electricity from wind power, and the technology continues to evolve and improve.

## Credit Cards

The concept of using credit to make purchases dates back hundreds of years, but the modern credit card as we know it today was invented by a man named Frank McNamara, along with Ralph Schneider and Matty Simmons, in 1950.

McNamara, who was a successful businessman in New York City, came up with the idea for a small, plastic card that people could use to make purchases at restaurants and other businesses without having to carry cash. He called it the "Diners Club" card and it quickly caught on, with thousands of people signing up within the first year.

Over time, other companies began to offer credit cards as well, and the credit card industry grew into the multi-billion dollar industry it is today. Today, there are a wide variety of credit cards available, each with its own unique features and benefits.

### Toll-Free

The concept of tolls, which is a fee charged to use a particular road or bridge, has been around for centuries, but the use of toll cards is a more recent development.

It is unclear which country had the very first toll card. However, one of the earliest examples of a toll card system was implemented in France in the mid-19th century. The French toll card was a paper card that allowed drivers to pre-pay for a certain number of trips across a particular toll road.

In the modern era, the first electronic toll card system was implemented in Norway in 1986. The system, called AutoPASS, uses a windshield-mounted transponder that automatically deducts tolls as a vehicle passes through a toll plaza.

Today, toll card systems are used in various countries around the world, with different names and technologies, such as the ETC system in Japan, the E-ZPass in the United States, and the Telepass in Italy.

## Tea

The exact origin of tea is unclear, but it is believed to have been first discovered and used as a beverage in China more than 4,000 years ago. According to legend, the Emperor Shen Nong was boiling water in his garden when some leaves from a nearby Camellia sinensis plant fell into the pot. The resulting infusion was refreshing and invigorating, leading to the discovery of tea as a beverage.

Tea became an important part of Chinese culture and was used for medicinal purposes as well as for social and ceremonial occasions. It was later introduced to other parts of the world, including Japan, Korea, and Southeast Asia, where it also became an integral part of the local cultures.

Tea was first introduced to Europe in the 16th century by Portuguese and Dutch traders, and it became a popular beverage among the aristocracy. The British East India Company began importing tea from China in the 17th century, leading to a surge in popularity and the establishment of tea plantations in India and Sri Lanka.

Today, tea is consumed around the world in many different forms and varieties, and it remains an important part of many cultures and traditions.

## Gyms

The concept of going to the gym for fitness and exercise purposes can be traced back to ancient Greece, where physical fitness was an important aspect of daily life. Gymnasiums were established in ancient Greece for the training and conditioning of athletes, as well as for military training. These gyms were typically outdoor spaces that included various equipment, such as weights and wrestling rings.

However, the modern concept of going to the gym as a recreational activity for health and fitness purposes can be traced back to the late 19th century. One of the early pioneers of this concept was a man named Gustav Zander, a Swedish physician who developed a system of exercise machines that were specifically designed for the purpose of improving health and fitness.

Zander's exercise machines were based on the principles of resistance training, and were designed to target specific muscle groups. He opened a gym in Stockholm in the 1860s that featured these machines, and the concept of "working out" for health and fitness purposes began to gain popularity.

In the early 20th century, other individuals and organizations began to promote the concept of going to the gym for health and fitness purposes. One of the most influential of these was Jack LaLanne, who opened the first modern health club in Oakland, California in the 1930s. LaLanne became known as the "godfather of fitness," and his health clubs helped to popularize the concept of going to the gym for exercise and fitness purposes.

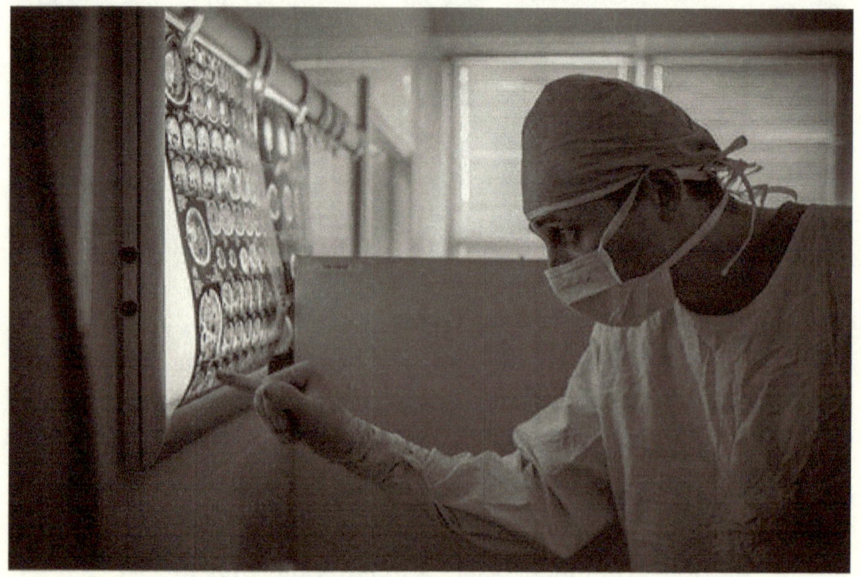

## CAT Scan

The computed tomography (CT) scan, also known as the CAT scan (Computed Axial Tomography), was invented independently by two researchers in the early 1970s.

One was British engineer Godfrey Hounsfield, who worked at EMI (Electric and Musical Industries) Laboratories in England. He developed the idea of using X-rays to create cross-sectional images of the human body and built the first CT scanner, which was used to image a brain in 1971. For this achievement, Hounsfield was awarded the Nobel Prize in Physiology or Medicine in 1979.

The other researcher was American physicist Allan Cormack of South African descent, who worked at Tufts University in the United States. Cormack independently developed the mathematical principles behind the technology, which enabled the reconstruction of 3D images from X-ray data. Cormack and Hounsfield are both credited with the invention of the CAT scan.

### Bicycles

The modern bicycle, as we know it today, was developed in the mid-19th century. There were many inventors who contributed to the development of the bicycle, and its evolution was a gradual process that took place over many years.

The first verifiable claim for a practically used bicycle belongs to German Baron Karl von Drais, who is credited with inventing the "running machine" in 1817. This was a wooden contraption with two wheels and a handlebar, which riders propelled by pushing their feet against the ground.

In the 1860s, Frenchman Pierre Michaux and his son Ernest began to develop a bicycle with pedals, which they called the "boneshaker" due to its rough ride. This was the first bicycle to be mass-produced and sold to the public.

Over the years, many inventors made significant improvements to the design of the bicycle, including the development of chain drives, pneumatic tires, and the modern diamond frame. However, Michaux's boneshaker is generally considered to be the first practical bicycle with pedals, which laid the foundation for the modern bicycle as we know it today.

**Zorbing**

Zorbing involves rolling down a hill or on a flat surface inside a large, transparent ball. It has become a popular recreational activity and has been used for team-building exercises and marketing campaigns.

The sport of zorbing, which involves rolling down a hill inside a large, transparent plastic sphere, was invented in the 1990s in New Zealand by Andrew Akers and Dwane van der Sluis. The two men were looking for an innovative way to have fun on their farm, and they came up with the idea of rolling down a hill inside a giant, inflatable ball. They called their invention the "Zorb," and it quickly became a popular attraction for thrill-seekers around the world. Today, there are many variations of zorbing, including water zorbing, where the ball is filled with water, and harness zorbing, where the rider is strapped into the ball.

## The Mop

The mop, as we know it today, was invented by Jacob Howe in 1837. However, there were earlier versions of cleaning tools that used a mop-like mechanism to clean floors. For example, ancient civilizations used brooms made from natural materials like twigs and leaves, which could be dampened and used to clean floors.

In the early 18th century, a British inventor named Jonathan Swift designed a mop-like cleaning device with a handle and a head made from strips of fabric. This was followed by a patent issued to Thomas Warren in 1834 for a "new and useful improvement in mop heads." However, it was Jacob Howe's design in 1837 that became the basis for the modern mop, featuring a wooden handle and an absorbent head made from yarn.

**The Mop:** The mop was invented by Manuel Jalón Corominas, a Spanish engineer, in 1956. He designed a mop that could be easily wrung out by using a lever, making it more efficient and easier to use than traditional mops.

## The Jet Boat

The credit for inventing the jet boat is usually attributed to Sir William Hamilton, a New Zealand engineer and inventor. In 1954, Hamilton developed a water jet propulsion system that allowed boats to move through shallow water, rapids, and other challenging environments where traditional propellers would be damaged or ineffective.

Hamilton's jet boat design featured a water intake at the bottom of the boat that drew water into a pump, which then expelled it through a nozzle at the stern, creating a powerful stream that propelled the boat forward. This innovative technology quickly became popular with commercial and recreational boaters, and today, jet boats are widely used for a variety of purposes, including racing, fishing, and tourism. This invention has become a popular recreational vehicle and has been used in a variety of industries, including tourism and search and rescue.

### Champagne

Champagne is a sparkling wine that is traditionally produced in the Champagne region of France. While there is some dispute over who exactly "invented" champagne, the process of creating sparkling wine can be traced back to the 16th century, when winemakers in the Champagne region began experimenting with different methods of making wine.

One important figure in the history of Champagne is a Benedictine monk named Dom Pérignon, who lived in the late 17th and early 18th centuries. While he did not invent champagne per se, he is often credited with improving the production process and helping to establish the reputation of Champagne as a high-quality sparkling wine.

Dom Pérignon is said to have introduced several innovations to the process of making Champagne, including the use of thicker glass bottles to prevent them from exploding during the secondary fermentation process, and the use of cork stoppers to seal the bottles. He also helped to refine the blending process, which involves mixing different grape varieties to create a unique flavor profile.

Over time, Champagne became increasingly popular among the French aristocracy, and by the 19th century it had become one of the most sought-after wines in the world. Today, Champagne is still considered one of the finest and most prestigious types of sparkling wine, and is enjoyed by wine lovers around the globe.

**Alcohol**

It is difficult to determine who specifically "invented" alcohol, as the process of fermentation that produces alcohol occurs naturally and has been utilized by various cultures for thousands of years.

Archaeological evidence suggests that alcoholic beverages were being produced as early as 7000-6600 BCE in China, and as early as 6000 BCE in the Near East. It is believed that early humans may have discovered the effects of fermentation by observing the natural process that occurs when fruits and grains are left to rot.

Over time, various cultures developed techniques for intentionally fermenting different types of fruits, grains, and other substances to create alcoholic beverages. For example, ancient Egyptians brewed beer using barley and other grains, while the Greeks and Romans produced wine using grapes.

So while it is difficult to pinpoint a single individual who "invented" alcohol, we do know that the use of alcoholic beverages has been a part of human history for thousands of years and has been refined and developed by many different cultures throughout that time.

## Tobacco

Tobacco is a plant that has been used for thousands of years by indigenous peoples in the Americas for medicinal, ceremonial, and recreational purposes. It is not known who discovered tobacco, as it has been used by various indigenous groups for centuries before the arrival of Europeans.

However, the widespread cultivation and use of tobacco in Europe and other parts of the world can be attributed to Christopher Columbus, who brought tobacco back to Europe from his first voyage to the Americas in 1492. From there, tobacco use spread rapidly throughout the world, becoming a major commodity in international trade and a popular recreational and social activity.

## Firearms

The invention of guns was a gradual process that involved the contributions of many people over several centuries. The earliest known device that used the explosive power of gunpowder was a Chinese invention called a "fire lance" or "fire spear," which was essentially a tube of bamboo or metal filled with gunpowder and shrapnel that was lit with a fuse and used as a weapon in battle. The first recorded use of the fire lance dates back to the 9th century.

Over time, the fire lance evolved into more sophisticated firearms, such as the hand cannon, which appeared in China in the 13th century and spread to Europe by the 14th century. The development of the matchlock mechanism in the 15th century allowed for more accurate and efficient firing of firearms, and subsequent innovations such as the flintlock and percussion cap further improved the reliability and firepower of guns.

So, while it's difficult to attribute the invention of guns to a single individual, it is safe to say that the development of firearms was the result of the cumulative efforts of many inventors and innovators over many centuries.

### The Glock

The Glock handgun was invented by an Austrian engineer named Gaston Glock. In the early 1980s, the Austrian military issued a request for a new handgun to replace its aging Walther P38 pistols. Glock, who had no prior experience in the firearms industry, responded to the request by designing a revolutionary new handgun that incorporated many innovative features.

Glock's design featured a polymer frame, which was lightweight, durable, and inexpensive to manufacture. The handgun also had a unique "safe action" system that combined a trigger safety, firing pin safety, and drop safety to prevent accidental discharges. In addition, the Glock had a high-capacity magazine and a low-profile design that made it easy to carry and conceal.

After extensive testing, the Austrian military selected Glock's handgun as its new service pistol, and the company quickly gained a reputation for producing reliable and innovative firearms. Today, Glock handguns are used by military, law enforcement, and civilian shooters around the world, and the company has become one of the most well-known and respected firearms manufacturers in the world.

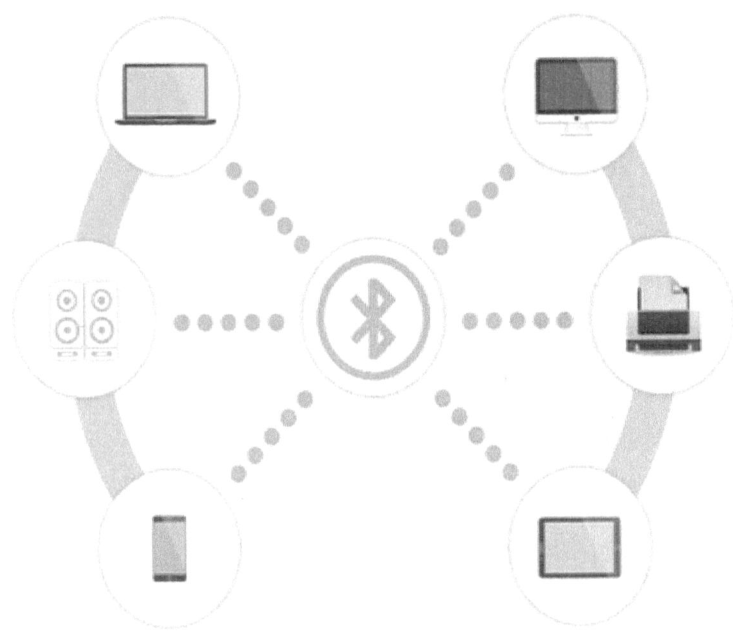

## Bluetooth

Bluetooth technology was invented by a group of engineers at Ericsson, a Swedish telecommunications company, in 1994. The group, led by Jaap Haartsen, developed a way to use short-range radio waves to connect devices wirelessly. The name "Bluetooth" was inspired by Harald Bluetooth, a 10th-century Danish king who united Denmark and

Norway, just as the technology was designed to unite different devices and industries. Today, Bluetooth technology is widely used in a variety of devices, from smartphones and laptops to wireless headphones and speakers.

## Laptops

The invention of laptops involved the contributions of several inventors and engineers over time, and it is difficult to attribute the invention to any one person. However, some key milestones in the development of laptops include:

- The first portable computer, the IBM 5100, was released in 1975. It weighed around 50 pounds and was more of a portable desktop computer than a true laptop.

- The first true laptop, the GRiD Compass, was released in 1982 by Grid Systems Corporation. It was designed by Bill Moggridge and weighed around 10 pounds.

• The first commercially successful laptop was the Toshiba T1100, released in 1985. It was designed by a team led by Tetsuya Mizoguchi at Toshiba and was one of the first laptops to use an Intel processor.

Since then, laptops have continued to evolve and improve with advances in technology and design. Today, laptops are essential tools for work, education, and entertainment, and they are used by millions of people around the world.

### Smartphones

The development of smartphones was the result of contributions from many people and companies over several decades. However, the first smartphone as we know it today can be traced back to IBM's Simon Personal Communicator, which was introduced in 1993.

The Simon had a touch screen, could send and receive faxes and emails and had basic PDA features such as a calendar, address book, and notes. However, it was not widely adopted and did not have cellular capabilities.

The first commercially successful smartphone was the BlackBerry, introduced in 1999. It was designed by Canadian company Research In Motion (now known as BlackBerry Limited) and was initially popular with businesspeople for its ability to send and receive emails on the go.

In 2007, Apple revolutionized the smartphone industry with the release of the iPhone, which introduced a touch screen and a user-friendly interface. The iPhone quickly became the most popular smartphone in the world, and its success inspired competitors to develop their own touch screen smartphones.

So, while no single person can be credited with inventing smartphones, many individuals and companies played important roles in their development over time.

**Algebra**

The concept of algebra has been developed and refined over thousands of years by many mathematicians from different cultures. However, the origins of algebra as a formal mathematical discipline can be traced back to the ancient Babylonians, who used algebraic methods to solve mathematical problems related to commerce, taxation, and trade.

In the Islamic Golden Age, the Persian mathematician Al-Khwarizmi is widely regarded as the "father of algebra" for his influential book "Al-Kitāb al-mukhtaṣar fī ḥisāb al-jabr wa-l-muqābala" (The Compendious Book on Calculation by Completion and Balancing), which introduced the systematic use of algebraic notation and techniques for solving linear and quadratic equations.

Algebra continued to evolve over the centuries, with major contributions from mathematicians such as René Descartes, Leonhard Euler, and Carl Friedrich Gauss, among others. Today, algebra is a fundamental area of mathematics with a wide range of applications in science, engineering, finance, and many other fields.

| Group→ / ↓Period | 1 | 2 | 3 | 4 | 5 | 6 | 7 | 8 | 9 | 10 | 11 | 12 | 13 | 14 | 15 | 16 | 17 | 18 |
|---|---|---|---|---|---|---|---|---|---|---|---|---|---|---|---|---|---|---|
| 1 | 1 H | | | | | | | | | | | | | | | | | 2 He |
| 2 | 3 Li | 4 Be | | | | | | | | | | | 5 B | 6 C | 7 N | 8 O | 9 F | 10 Ne |
| 3 | 11 Na | 12 Mg | | | | | | | | | | | 13 Al | 14 Si | 15 P | 16 S | 17 Cl | 18 Ar |
| 4 | 19 K | 20 Ca | 21 Sc | 22 Ti | 23 V | 24 Cr | 25 Mn | 26 Fe | 27 Co | 28 Ni | 29 Cu | 30 Zn | 31 Ga | 32 Ge | 33 As | 34 Se | 35 Br | 36 Kr |
| 5 | 37 Rb | 38 Sr | 39 Y | 40 Zr | 41 Nb | 42 Mo | 43 Tc | 44 Ru | 45 Rh | 46 Pd | 47 Ag | 48 Cd | 49 In | 50 Sn | 51 Sb | 52 Te | 53 I | 54 Xe |
| 6 | 55 Cs | 56 Ba | | 72 Hf | 73 Ta | 74 W | 75 Re | 76 Os | 77 Ir | 78 Pt | 79 Au | 80 Hg | 81 Tl | 82 Pb | 83 Bi | 84 Po | 85 At | 86 Rn |
| 7 | 87 Fr | 88 Ra | | 104 Rf | 105 Db | 106 Sg | 107 Bh | 108 Hs | 109 Mt | 110 Ds | 111 Rg | 112 Cn | 113 Uut | 114 Fl | 115 Uup | 116 Lv | 117 Uus | 118 Uuo |

| Lanthanides | 57 La | 58 Ce | 59 Pr | 60 Nd | 61 Pm | 62 Sm | 63 Eu | 64 Gd | 65 Tb | 66 Dy | 67 Ho | 68 Er | 69 Tm | 70 Yb | 71 Lu |
|---|---|---|---|---|---|---|---|---|---|---|---|---|---|---|---|
| Actinides | 89 Ac | 90 Th | 91 Pa | 92 U | 93 Np | 94 Pu | 95 Am | 96 Cm | 97 Bk | 98 Cf | 99 Es | 100 Fm | 101 Md | 102 No | 103 Lr |

## The Periodic Table

The periodic table was not invented by a single person, but rather was the result of contributions from many scientists over a period of several decades. However, the person credited with creating the first version of the periodic table as we know it today is Dmitri Mendeleev, a Russian chemist.

In 1869, Mendeleev published a paper in which he arranged the known elements in a table based on their properties and atomic weights. He left gaps in the table where he predicted that new elements would be discovered in the future, and he also used the table to make predictions about the properties of these as-yet-undiscovered elements.

Mendeleev's periodic table was a significant breakthrough in the field of chemistry, and it became widely used in scientific research and education. Over time, the periodic table was refined and expanded as new elements were discovered and our understanding of atomic structure and properties improved. Today, the periodic table is a fundamental tool in chemistry and is used to understand the behavior of elements and their interactions with other substances.

### The Rubik cube

The Rubik's Cube was invented by Ernő Rubik, a Hungarian sculptor and professor of architecture, in 1974. Rubik created the puzzle as a way to teach his students about three-dimensional geometry and spatial relationships. Initially called the "Magic Cube," Rubik's invention became popular in Hungary and eventually spread worldwide after it was licensed to the Ideal Toy Corp in 1980 and renamed the "Rubik's Cube." The puzzle became a global phenomenon and remains a popular toy and puzzle to this day.

### Drones

The microwave oven was not invented by any one person, but rather it was the result of many scientists and inventors working on various technologies related to microwaves and cooking.

In 1945, Percy Spencer, an engineer working for the Raytheon Corporation, was conducting experiments with radar technology when he noticed that a candy bar in his pocket had melted. Intrigued, he decided to experiment with food and eventually developed the first microwave oven, which was called the "Radarange." Raytheon filed a patent for the invention in 1945, and the first commercial microwave ovens were sold in 1947.

However, the development of the microwave oven was not solely the work of Percy Spencer or Raytheon. Many other inventors and scientists had been working on related technologies for decades, including the development of magnetrons and radar technology during World War II.

## Newspapers

The concept of newspapers or written news can be traced back to ancient Rome where official notices and announcements called Acta Diurna were posted in public places. However, the modern newspaper as we know it today was developed in Europe in the 17th century.

The first regularly published newspaper is credited to Johann Carolus, a German publisher who in 1605 began publishing a weekly news bulletin in Strasbourg, France called Relation aller Fürnemmen und gedenckwürdigen Historien (Account of all distinguished and memorable news).

Other early newspapers include the Oxford Gazette (later the London Gazette) in England, which began publishing in 1665 and focused on official government news, and the Haarlems Dagblad, a

Dutch newspaper which began in 1656 and was the first to feature advertisements.

Newspapers evolved and became more widespread in the following centuries, with many different newspapers being published in various countries around the world.

## Disposable Masks

The invention of disposable masks is often attributed to Sara Little Turnbull, who in 1958 was working as a textile engineer for the 3M company. Turnbull was tasked with developing a new type of surgical mask that would be more effective, comfortable, and convenient than the cloth masks that were commonly used at the time.

Turnbull and her team came up with a design for a lightweight, disposable mask made from non-woven synthetic fabric. The mask had a metal strip over the nose to ensure a secure fit and was held in place by elastic bands that looped over the ears. This design was patented in 1961 and became the prototype for the modern disposable mask.

Although Turnbull is often credited with inventing the disposable mask, it's worth noting that other researchers and companies were also working on similar designs around the same time. However, the 3M mask was the first to gain widespread use in the medical community, and it set the standard for disposable masks that are still used today.

### The Wristwatch

The first wristwatch was invented by Patek Philippe in 1868, and it was primarily designed for women. However, it was not until the early 20th century that wristwatches became popular and widely used, particularly by men during World War I. The first watches were developed in the 16th century and the exact inventor is not known. However, one of the earliest and most important pioneers of watchmaking was Peter Henlein, a German locksmith and clockmaker who is credited with inventing the portable watch in the early 16th century. Henlein's watches were small, egg-shaped devices that were worn as pendants and were powered by small springs. While the first watches were not very accurate and needed to be wound frequently, they paved the way for the development of more sophisticated timepieces in the centuries that followed.

### Clowns

The concept of clowns, or comedic performers who use physical humor, exaggerated gestures, and costumes to entertain audiences, has been around for centuries, if not millennia. The ancient Greeks and Romans had their own forms of clownish characters in theater and festivals, and similar figures appear in cultures around the world.

However, the modern image of the clown, with its distinctive makeup, colorful clothing, and oversized shoes, can be traced back to the circus traditions of the 19th century. The first circus clown to use the now-iconic whiteface makeup and exaggerated features was Joseph Grimaldi, a British performer who rose to fame in the early 1800s.

Grimaldi's innovations helped define the modern clown archetype and inspired generations of performers who followed in his footsteps.

Today, clowns continue to be a popular form of entertainment, appearing in circuses, parades, children's parties, and other events around the world.

## Banks

The concept of banking and financial institutions can be traced back to ancient times. The earliest known banks were established in ancient Babylon around 2000 BCE, where temples and palaces provided safe storage for valuables such as gold and silver.

However, the modern banking system as we know it today was developed during the Renaissance period in Italy. In the 14th century, the Medici family of Florence, Italy, established the first modern banking system, which included the use of bills of exchange, letters of credit, and other financial instruments.

Over time, banking systems evolved and became more complex, with the establishment of central banks, investment banks, and commercial banks. Today, banks play a crucial role in the global economy, providing individuals and businesses with access to credit, financing, and other financial services.

## The ATM

The first ATM (Automated Teller Machine) was invented by John Shepherd-Barron, a Scottish inventor and businessman. He came up with the idea of a machine that could dispense cash to customers outside of bank hours and on weekends, while he was in the bath, and he subsequently developed and installed the first machine in a branch of Barclays Bank in Enfield, north London, in 1967.

There have been other claims that the ATM was invented by other individuals or companies, but Shepherd-Barron is widely credited as the inventor of the first modern ATM. Automatic Teller Machine (ATM) - Invented by Luther George Simjian in 1939, but first introduced in Japan in the 1970s.

**Pajamas**

The origins of pajamas are not entirely clear, as various cultures around the world have worn different types of sleepwear for centuries. However, the modern-day pajamas that we know today were popularized by British colonizers who were exposed to the traditional Indian garment known as the "paejama" during their travels to India in the 18th and 19th centuries.

The British adapted the Indian paejama into a more comfortable and practical version that consisted of a two-piece set of loose-fitting trousers and a shirt. These early versions of pajamas were typically made of silk or cotton and were primarily worn by men as loungewear.

Over time, the popularity of pajamas spread throughout Western cultures, and they became increasingly accessible to people of all genders and ages. Today, pajamas are a staple of comfortable and casual attire and are worn by millions of people around the world. While the specific inventor of pajamas is not known, their development can be traced back to the influence of various cultures and traditions over time.

## Neckties

The origin of neckties can be traced back to the 17th century in Europe. However, the modern necktie as we know it today was developed in the 19th century. While there is no one inventor of the necktie, the evolution of men's fashion and the necktie was a collaborative effort over time.

The precursor to the modern necktie was the cravat, a type of neckcloth worn by Croatian mercenaries in the 17th century. The French King Louis XIV was impressed with the cravat and made it fashionable in France. In the 19th century, the cravat was replaced by the four-in-hand necktie, which is the ancestor of today's necktie.

The four-in-hand necktie was named after the four-in-hand carriage, a popular mode of transportation at the time. It was long and narrow, with pointed ends, and was tied in a knot similar to the modern-day knot. The knot was originally called the "four-in-hand knot," but over time, the term "necktie" came into use.

Throughout the 19th and 20th centuries, neckties continued to evolve in terms of style, fabric, and knotting techniques. Today, neckties come in a wide range of colors, patterns, and materials, and are an essential part of men's formal and business attire. While there is no one person who can be credited with inventing the necktie, it is the result of centuries of evolution and innovation in men's fashion.

**The Braille**

The Braille writing system was invented by Louis Braille, a French educator who was born in Coupvray, France in 1809. Braille became blind at a young age due to an accident and was educated at the Royal Institution for Blind Youth in Paris, where he excelled as a student.

In 1824, Braille began working on a system of raised dots that blind people could use to read and write. He based his system on a military code called "night writing" that used raised dots and dashes to communicate silently in the dark. Braille simplified the night writing system and developed a system of six dots that could represent the letters of the alphabet, numbers, and punctuation marks.

Braille's system was initially met with skepticism and resistance, but it gradually gained acceptance and spread throughout the world. Today, the Braille system is used by millions of blind and visually impaired people to read and write in many different languages.

**Backpacks**

As mentioned earlier, the modern backpack, also known as a knapsack, is attributed to the Germans. However, the concept of carrying things on one's back dates back thousands of years.

In ancient times, people used simple bags or pouches made of animal skins, leather, or cloth to carry their belongings. In some cultures, people used baskets or other containers to carry goods on their backs.

The first known use of a backpack in a military context was by the Roman Legionnaires. They used a type of backpack called a "military rucksack" to carry their gear and supplies during long marches.

In the 19th century, backpacks were used by trappers, prospectors, and miners to carry their equipment and supplies. The first commercially successful backpack was created in 1938 by Gerry Cunningham, who founded the company Gerry Outdoors. His backpack design was inspired by the packboards used by the US Forest Service to carry equipment.

However, it was the German company Deuter that popularized the modern backpack design, which includes shoulder straps and a compartment for storage. Deuter began manufacturing backpacks for hiking and mountaineering in the early 20th century, and their design became very popular. Today, there are many different types of backpacks, and they are used for a variety of purposes, from hiking and camping to school and work.

## The Mechanical Pencil

The modern mechanical pencil, which uses a lead refill system and a mechanism to advance the lead, was invented by an American named Charles R. Keeran in 1879. Keeran's design was later improved upon by other inventors, including Tokuji Hayakawa, the founder of Sharp Corporation, who developed the "Ever-Sharp" mechanical pencil in 1915. The Ever-Sharp pencil used a twistable mechanism to advance the lead, which is still a common design feature in mechanical pencils today.

## Paper

The invention of paper is attributed to Cai Lun, a Chinese eunuch and court official who lived during the Eastern Han Dynasty (25-220 CE). In 105 CE, Cai Lun developed a method for making paper from mulberry bark, hemp, old rags, and other plant materials. This new method of papermaking revolutionized the world of writing and record-keeping, and paper eventually replaced other writing materials such as bamboo, silk, and animal skins. Papermaking techniques later spread to other parts of the world, including the Islamic world, where it was further developed, and Europe, where it arrived in the 12th century.

### Scissors

The invention of scissors dates back to ancient times, so it is difficult to identify a single individual who can be credited with their invention. The first scissors were likely developed in ancient Egypt or Mesopotamia around 1500 BCE, and were made from bronze.

Over time, scissors evolved and were made from different materials such as iron and steel. Different cultures and regions also developed their own types of scissors, such as the spring scissors of China and the cross-blade scissors of Europe.

In terms of a specific inventor of modern scissors, the credit is often given to Robert Hinchliffe, a British manufacturer who patented the first modern scissors in 1761. However, it is important to note that Hinchliffe's scissors were an improvement on existing designs, and he was not the first person to create a pair of scissors.

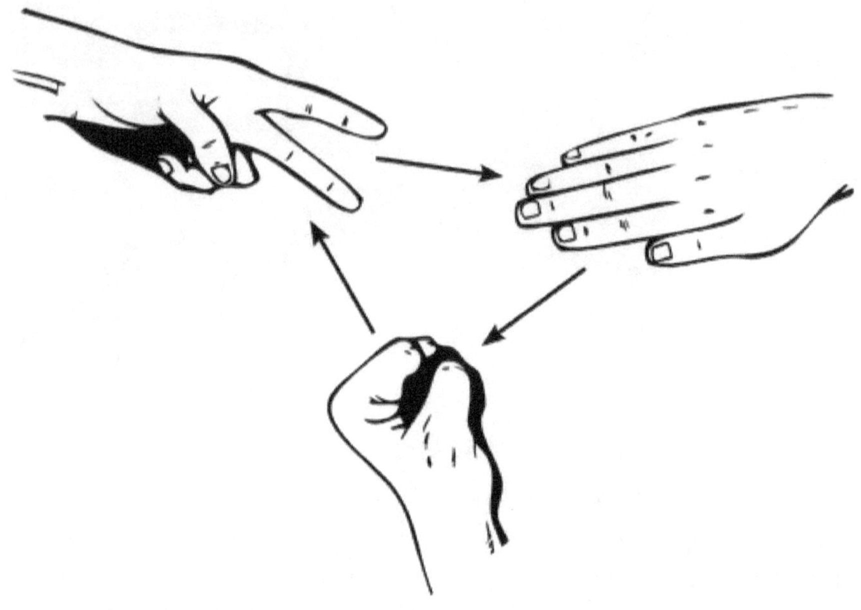

### Rock Paper Scissors

The origin of the game rock-paper-scissors (also known as roshambo) is not entirely clear, as variations of the game have been played in different cultures throughout history. However, it is believed to have originated in China around 2,000 years ago as a hand game called "shoushiling." The game was later introduced to Japan, where it became popular under the name "jan-ken."

The modern version of the game, which uses the symbols of rock, paper, and scissors, is said to have originated in Japan in the 19th century. It became widely popular in Japan in the early 20th century, and later spread to other parts of the world.

It is unclear who specifically invented the modern version of the game, as it is believed to have developed over time through cultural exchange and evolution. Nevertheless, rock-paper-scissors remains a popular and widely recognized game around the world.

### Sneakers

The history of sneakers can be traced back to the late 18th century when people began using rubber soled shoes for sporting activities. However, the modern sneaker as we know it today was first invented in 1917 by the U.S. Rubber Company, which later became known as Keds.

Keds was the first company to mass-produce sneakers with rubber soles, and they quickly became popular among athletes and the general public. The term "sneaker" was coined because the rubber sole made the shoe virtually silent, allowing people to "sneak" up on others.

Since then, many other companies have developed their own versions of sneakers, and they have become a staple of fashion and everyday footwear for people around the world.

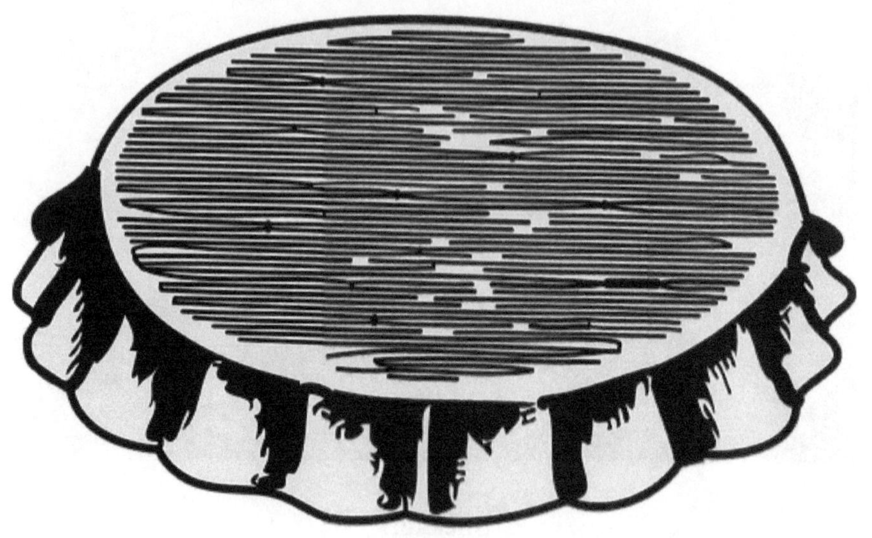

## Soda

The history of soda is quite complex, and it is difficult to pinpoint a single inventor. The concept of carbonated water, which is the main ingredient in soda, dates back to the 18th century when natural springs containing carbon dioxide were discovered in various parts of Europe. People would travel to these springs to drink the carbonated water for its supposed health benefits.

In the late 18th century, Joseph Priestley, an English chemist, discovered a method of infusing water with carbon dioxide by using a fermentation process. He called the resulting drink "impregnated water" and noted its refreshing taste.

In the early 19th century, John Mervin Nooth, an Englishman, patented a method of producing carbonated water by dissolving carbon dioxide in water under pressure. This method made carbonated water more widely available and paved the way for the creation of soda.

However, the first commercial soda was created in 1807 by Benjamin Silliman, a Yale University chemistry professor. He produced flavored carbonated water that he called "soda water" and sold it to the public. The drink became popular, and other companies began to produce their own versions of soda.

So while there isn't a single inventor of soda, several people played important roles in its development and popularity.

### The Microwave

The microwave oven was not invented by any one person, but rather it was the result of many scientists and inventors working on various technologies related to microwaves and cooking.

In 1945, Percy Spencer, an engineer working for the Raytheon Corporation, was conducting experiments with radar technology when he noticed that a candy bar in his pocket had melted. Intrigued, he decided to experiment with food and eventually developed the first microwave oven, which was called the "Radarange." Raytheon filed a patent for the invention in 1945, and the first commercial microwave ovens were sold in 1947.

However, the development of the microwave oven was not solely the work of Percy Spencer or Raytheon. Many other inventors and scientists had been working on related technologies for decades, including the development of magnetrons and radar technology during World War II.

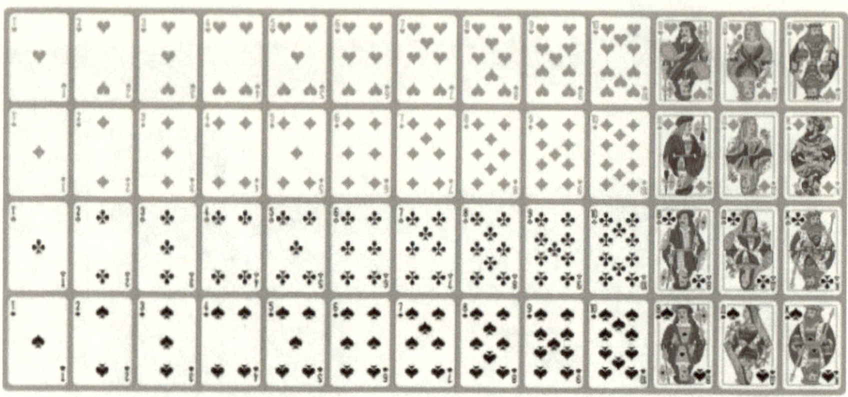

## Deck of Cards

The invention of playing cards is not attributed to a single person or culture as its origin is shrouded in mystery and speculation. However, playing cards are believed to have originated in China during the Tang dynasty (618-907 AD) and were used for gambling and other games. The earliest recorded references to playing cards in Europe date back to the 14th century, suggesting that the game spread from Asia to Europe through trade and cultural exchange.

It's worth noting that different regions have developed their own variations of playing cards, with unique suits, designs, and values. For example, the modern deck of 52 playing cards used in the Western world originated in France in the 15th century and features four suits: hearts, diamonds, clubs, and spades. In contrast, the traditional deck of Chinese playing cards consists of 32 cards, with four suits based on Chinese symbols and characters.

## Golf

The origins of golf are not entirely clear, as many cultures throughout history have played games that involve hitting a ball with a stick or club. However, the modern game of golf as it is known today is generally believed to have originated in Scotland in the 15th century.

There are several historical references to golf being played in Scotland as far back as the 1400s, and the first recorded game of golf took place in 1457 in the town of St. Andrews. The game spread throughout Scotland and eventually to England and other parts of the world, evolving over time into the sport that is played today.

While it is not possible to attribute the invention of golf to a single individual, many notable figures in the history of golf have contributed to the development and popularization of the sport over the centuries. Some of these include Old Tom Morris, who was a prominent Scottish golfer and course designer in the 1800s, and Harry Vardon, a professional golfer from the Channel Islands who was instrumental in the development of the modern golf swing.

## Baseball

The modern game of baseball is often credited to Abner Doubleday, who was said to have invented the game in Cooperstown, New York in 1839. However, there is no conclusive evidence to support this claim and most historians now believe that baseball evolved from earlier bat-and-ball games that were played in Europe and North America.

One of the earliest known versions of baseball was played in England in the 18th century and was called rounders. The game was brought to North America by English immigrants and was played in various forms throughout the 19th century.

The first recorded baseball game in North America took place in 1846 between the Knickerbocker Baseball Club and the New York Nine. Over time, the rules of the game were standardized and professional leagues were established. Today, baseball is one of the most popular sports in the United States and is played at all levels, from youth leagues to the major leagues.

### Tennis

The modern game of tennis was developed in the late 19th century by a number of individuals in England, but no one person can be credited with inventing the game.

One of the earliest versions of tennis was played in France in the 12th century, where players used their hands to hit a ball back and forth over a rope or net. Over time, the game evolved, and by the 16th century, rackets were used and the game was played on a court.

In the 1870s, the first tennis club was established in Leamington Spa, England, and the modern rules of the game were developed. Major Walter Clopton Wingfield is often credited with inventing the game of lawn tennis in 1873, and he patented the equipment and rules for the game in 1874. However, other individuals, such as Harry Gem and Augurio Perera, also made important contributions to the development of modern tennis.

Today, tennis is a popular sport played all over the world, and it has a rich history and tradition that has evolved over centuries.

### Pen

The ballpoint pen was invented by László Bíró, a Hungarian journalist, in 1938. He was looking for a pen that would not smudge or smear like the fountain pens of his day. Bíró noticed that the ink used in newspaper printing dried quickly and did not smear, so he came up with the idea of using a tiny ball bearing to transfer ink to paper. Bíró and his brother Georg worked on developing the ballpoint pen and eventually patented their design. The first commercially successful ballpoint pens were produced in 1945 by the Reynolds International Pen Company in the United States. The "Biro" Pen: László Bíró, a Hungarian-Argentine journalist, invented the ballpoint pen in 1938. However, it was improved by Brazilian inventor João Carlos de Camargo e Costa, who created the modern version in the 1940s.

## Internet

The internet was not invented by a single person, but rather it was developed over time by numerous individuals and organizations.

The precursor to the modern internet was the ARPANET, which was created by the United States Department of Defense's Advanced Research Projects Agency (ARPA) in the 1960s. The goal of the ARPANET was to create a decentralized, fault-tolerant communication network that could survive a nuclear attack.

Several key individuals played important roles in the development of the ARPANET and the internet, including computer scientists like Vint Cerf and Bob Kahn, who developed the Transmission Control Protocol (TCP) and the Internet Protocol (IP), which form the basis of the internet's communication protocols.

Other important contributors to the development of the internet include Tim Berners-Lee, who created the World Wide Web, and Sir Timothy John Berners-Lee, who created the first web browser.

Overall, the internet was created by a collaborative effort involving many individuals and organizations, and it continues to evolve and grow to this day.

## Video streaming

The development of video streaming technology can be attributed to multiple individuals and organizations over time, so it's difficult to identify a single inventor.

In the early days of video streaming, RealNetworks (formerly known as Progressive Networks) played a significant role in developing the technology. They released RealVideo in 1997, which was one of the first video streaming formats for the web.

Microsoft also played a role in the development of video streaming with their release of Windows Media Player in 1999, which included support for streaming media.

Other companies and individuals have contributed to the development of video streaming over the years, including Adobe (with their Flash Video format), Apple (with QuickTime and HTTP Live Streaming), and Netflix (with their proprietary streaming technology).

So while there is no one inventor of video streaming, it is the result of the contributions and innovations of many people and organizations over time.

## MP3

The development of the MP3 audio format was a collaborative effort involving several individuals and organizations.

The key people involved in the creation of the MP3 format are:

- Karlheinz Brandenburg, a German engineer who is often credited as the "father of MP3" for his work on developing the MP3 standard at the Fraunhofer Institute for Integrated Circuits in the late 1980s and early 1990s.

- Professor Dieter Seitzer, a colleague of Brandenburg's at the University of Erlangen-Nuremberg, who provided input on psychoacoustic principles and helped test and refine the MP3 technology.

- The Moving Picture Experts Group (MPEG), a committee of experts from around the world who developed the standard for compressing digital audio and video data, which included the MP3 format.

Therefore, the invention of the MP3 format was not the work of a single person, but rather the result of collaboration between several engineers and experts in the field of digital audio.

## Rubber Band

The modern rubber band, as we know it today, was invented in 1845 by the British inventor Stephen Perry. Perry was inspired to create the rubber band after receiving a shipment of vulcanized rubber from South America. He realized that the material had potential as a stretchy, flexible band that could be used for a variety of purposes.

Perry's rubber bands quickly became popular, and he founded a company called Messrs Perry and Co. to manufacture and sell them.

Today, rubber bands are used in a wide variety of applications, from holding together stacks of paper to launching homemade rockets.

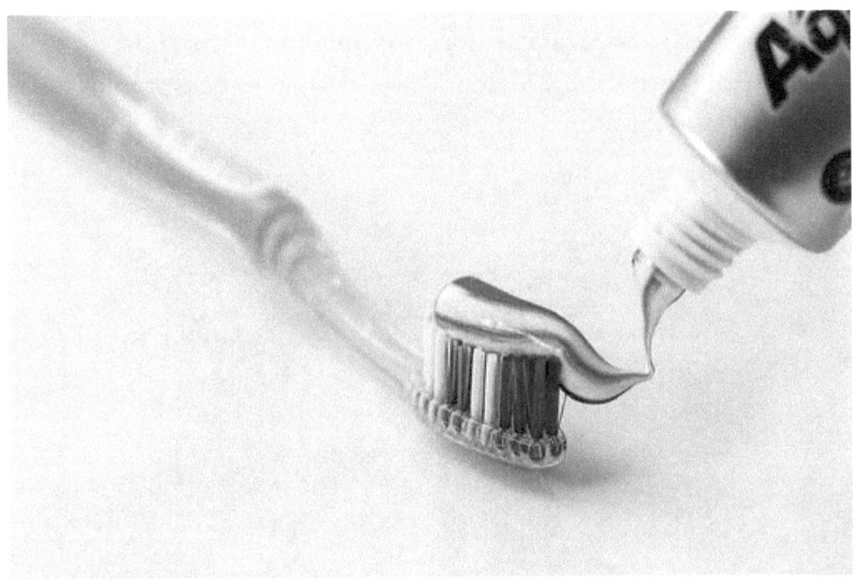

### Toothbrush

The concept of cleaning teeth with a brush dates back thousands of years to ancient civilizations such as the Egyptians, Greeks, and Romans. However, the modern toothbrush as we know it today was invented in the 18th century.

The first mass-produced toothbrush was made by William Addis, an English inventor, in 1780. Addis was serving time in prison at the time and reportedly came up with the idea for a toothbrush after seeing other prisoners using rags and soot to clean their teeth.

Addis carved a handle out of bone and drilled small holes in it to insert bristles made from the hair of cows. He later set up a toothbrush manufacturing business in England and began producing toothbrushes on a large scale.

Over the years, toothbrushes have continued to evolve and improve, with new materials, designs, and features. Today, toothbrushes come in a variety of shapes and sizes, with bristles made from nylon or other synthetic materials, and are an essential part of oral hygiene for people around the world. Dental professionals recommend brushing teeth twice a day for two minutes each time, using a soft-bristled toothbrush and fluoride toothpaste, to help prevent tooth decay and gum disease.

## Vacuum

The vacuum cleaner was invented by several people in the late 19th century, but the first practical and portable vacuum cleaner was developed by British inventor Hubert Cecil Booth in 1901.

Booth's vacuum cleaner used a motorized pump to create suction and a large, horse-drawn carriage to transport the equipment. The machine was large and expensive, but it was effective at removing dust and dirt from carpets and upholstery.

In 1907, an American inventor named James Murray Spangler developed the first portable vacuum cleaner with an electric motor. Spangler, who suffered from asthma, was looking for a way to clean his home without stirring up dust and aggravating his condition. He used a fan motor, a soapbox, and a pillowcase to create his first prototype, which he later refined and patented.

Spangler sold the rights to his invention to William Henry Hoover, who founded the Hoover Company and became one of the leading manufacturers of vacuum cleaners in the 20th century. Today, vacuum cleaners come in a variety of sizes and styles, with a range of features such as bagless technology, HEPA filters, and robotic operation. They are an essential tool for keeping homes and workplaces clean and healthy.

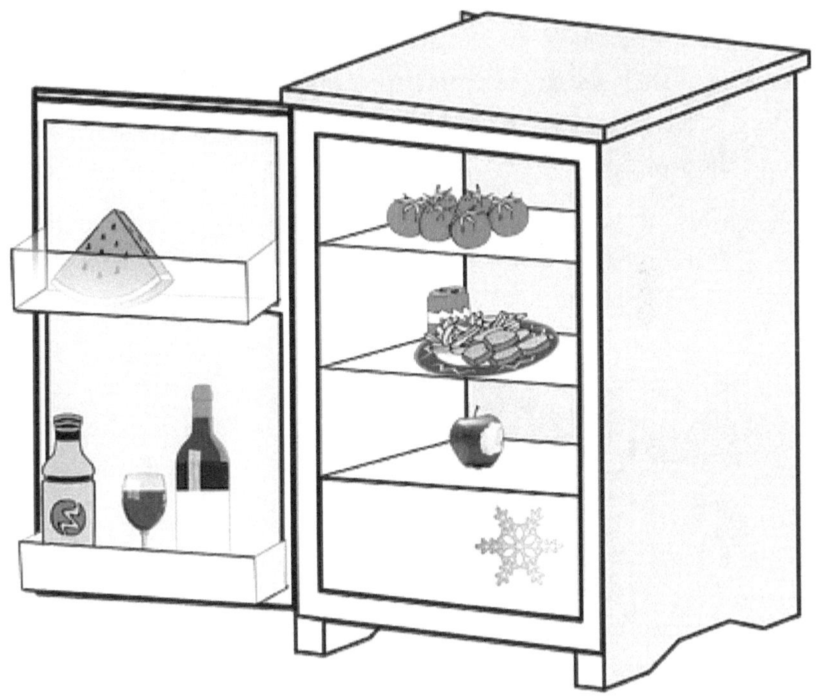

### Fridge

The modern refrigerator, as we know it today, was invented by several inventors and innovators over several decades.

The first commercially successful refrigerator was invented by an American engineer named Nathaniel B. Wales in 1876. Wales' refrigerator used a compressor and ammonia to produce a cooling effect, and it was the first refrigerator to be sold in large numbers.

In 1913, a French engineer named Marcel Audiffren developed the first domestic refrigerator that used a hermetically sealed compressor, which made it smaller, more efficient, and more affordable.

Other inventors and companies, such as Albert T. Marshall, General Electric, and Kelvinator, also played a role in the development and improvement of refrigerators throughout the 20th century.

Today, refrigerators come in a variety of styles and sizes, with a range of features such as ice makers, water dispensers, and smart technology. They are an essential appliance for preserving food and keeping it fresh, and are found in homes, businesses, and industries around the world.

**Blow dryer**

The first handheld hair dryer was invented by Alexander Godefroy in the 1890s. It was a large, seated device that used hot air to dry hair. However, the modern, portable hair dryer that we are familiar with today was invented by Gabriel Kazanjian in 1920. Kazanjian's invention was called the "blow dryer" and it was much smaller and more portable than Godefroy's device. Over the years, the blow dryer has undergone many improvements and variations, but Kazanjian is credited with its original invention.

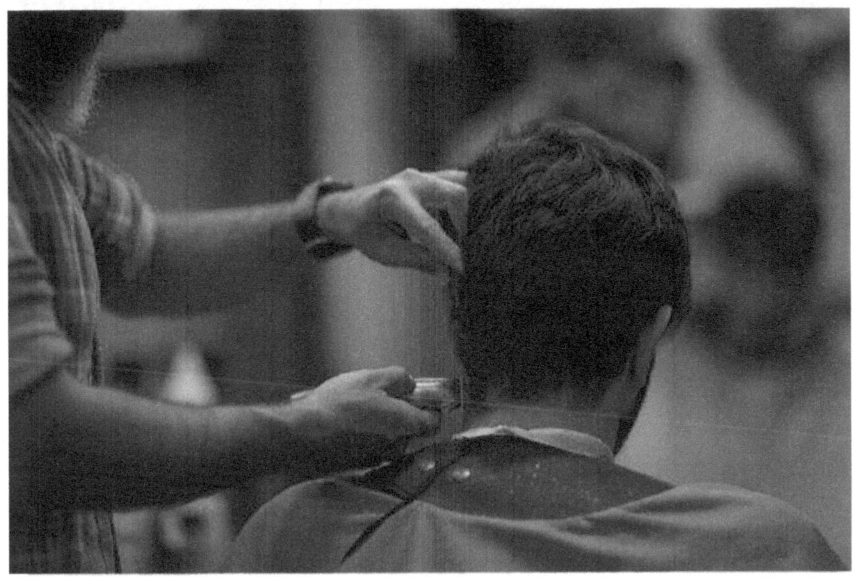

## Men's Electric Clippers

The first electric hair clipper for men was invented by a Serbian-American inventor named Nikola Bizumić in 1921. Bizumić's invention, which he called the "tondeuse électrique," was a revolutionary innovation in the field of men's grooming.

Prior to the invention of electric hair clippers, barbers had to use manual scissors or razors to cut and trim men's hair. This was a time-consuming and labor-intensive process that could be uncomfortable for the customer.

Bizumić's invention changed all of that. His electric hair clipper allowed barbers to cut hair quickly and efficiently, with minimal discomfort to the customer. Over time, electric hair clippers became a standard tool in barbershops and salons around the world.

## Chess

The origins of chess are unclear, but the game is believed to have originated in northern India or eastern Iran around the 6th century AD. The game then spread to Persia, Arabia, and eventually to Europe through trade and conquest.

There is no single person credited with inventing chess, as it evolved over time from earlier games such as chaturanga. The modern form of chess that is played today emerged in Europe during the 15th century.

The origins of chess are not entirely clear, but the game is believed to have originated in northern India or Afghanistan around the 6th century AD. The earliest predecessor of chess was a game called chaturanga, which means "four divisions" in Sanskrit, and was played with pieces representing the four branches of the Indian military: infantry, cavalry, elephants, and chariots.

The modern game of chess, as we know it today, evolved in Europe during the Middle Ages, and it is not attributed to any single individual. The rules and pieces of chess were gradually refined over time, and by the 15th century, the game had reached its current form. It is believed that modern chess pieces, with their symbolic shapes and names, were developed in Europe during the 15th century.

### Lighter

The first lighter was invented by Johann Wolfgang Dobereiner, a German chemist, in 1823. His lighter, called the Dobereiner's lamp, used a chemical reaction to produce a flame. However, the modern lighter that we are familiar with today, which uses a spark to ignite flammable gas, was invented in the early 1900s by a variety of inventors, including Carl Auer von Welsbach, Ignacy Łukasiewicz, and Theodor von Karman. The first commercial lighter, called the "Wonderful" lighter, was invented in 1910 by the Austrian company IMCO. Since then, many different types of lighters have been invented, including pocket lighters, torch lighters, and electronic lighters.

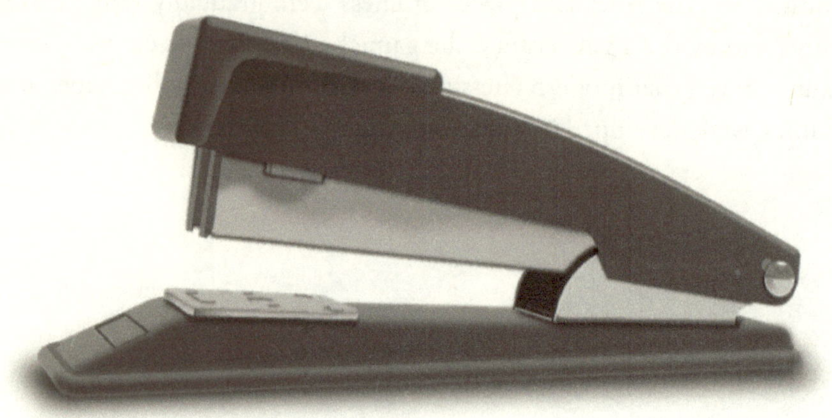

## Staplers

The first known stapler was made in the 18th century in France for King Louis XV. It was designed for the French court's clerks and was used to fasten their official documents with a tiny brass staple. However, the modern stapler as we know it was not invented until the late 19th century.

The first practical stapler was patented in 1877 by Henry R. Heyl, an American inventor. Heyl's stapler used a double-ended strip of staples that were held together with glue. The user would slide the strip into the stapler and press down on a lever, which pushed a blade through the strip and bent the ends of the staple to hold the papers together.

Since Heyl's invention, many different types of staplers have been developed, including manual and electric models, as well as different types of staples designed for specific uses.

The modern stapler, which uses metal staples to fasten papers together, was invented by the American inventor George W. McGill in 1877. His design was based on earlier machines that used brass fasteners, but he improved the design by using a smaller, more efficient staple. Over the years, various improvements and refinements have been made to the stapler, leading to the wide variety of stapler designs that we have today.

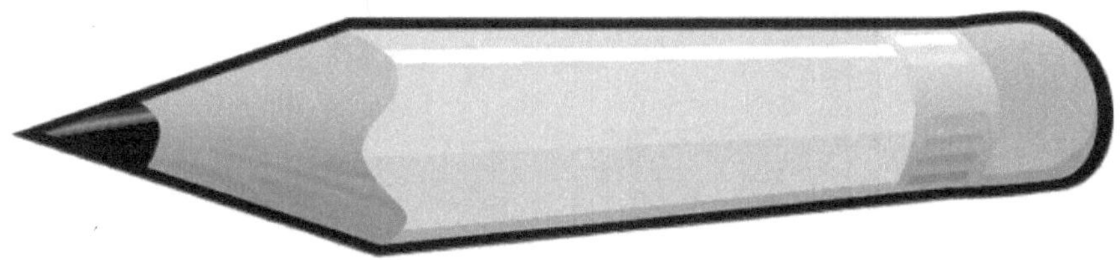

## Pencils

The modern pencil, as we know it today, was invented in the late 18th century by Nicholas-Jacques Conte, a French artist and scientist. Conte developed a method for producing high-quality graphite sticks by grinding graphite powder, mixing it with clay, and then firing it in a kiln. This made the graphite more durable and less prone to smudging. Conte's method is still used today to produce high-quality pencils. However, it's worth noting that there were earlier versions of pencils that were made using other materials, such as lead or charcoal, and some of this date back as far as ancient Greece and Rome.

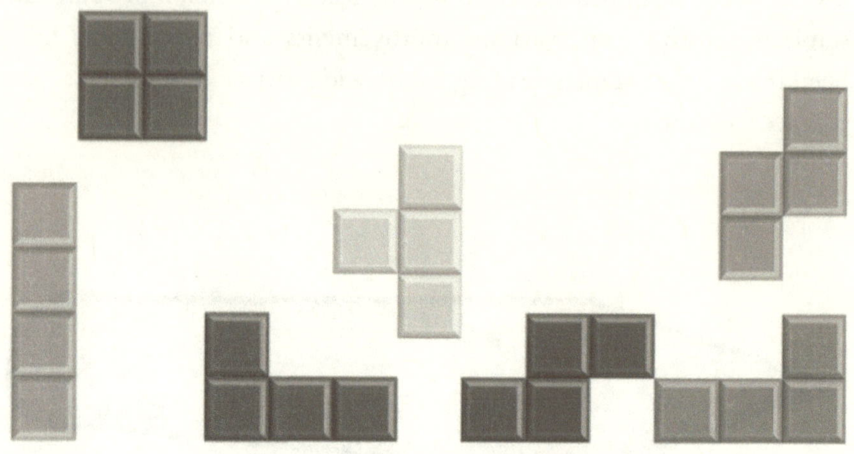

## Video Games

The first video game is generally credited to "Tennis for Two," which was created by physicist William Higinbotham in 1958. The game was played on an oscilloscope, and involved two players using knobs to control the movement of lines on the screen, which represented a tennis court and ball. While there were earlier interactive electronic games, such as the cathode-ray tube amusement device developed by Thomas T. Goldsmith Jr. and Estle Ray Mann in 1947, Tennis for Two is considered the first video game because it used a graphical display to depict the game action.

**Tetris: This popular puzzle game was created by Russian Alexey Pajitnov in 1984, and has been enjoyed by millions of people around the world.**

### Instant Ramen

Instant ramen was invented by Momofuku Ando, a Taiwanese-Japanese inventor and businessman. Ando first developed the concept of instant ramen in 1958 while working for his company, Nissin Foods. He came up with the idea after observing the long lines of people waiting for hot bowls of ramen during post-World War II Japan, and wanted to find a way to provide a quick and convenient alternative.

Ando spent over a year experimenting with different noodle recipes and methods of dehydrating and packaging them. He finally succeeded in creating a product that could be cooked in just a few minutes by adding boiling water. The first instant ramen product, Chicken Ramen, was launched in 1958 and became an instant success in Japan.

Since then, instant ramen has become a worldwide phenomenon, with countless brands and varieties available in supermarkets around the globe. Ando's innovation revolutionized the way people consume noodles and made them accessible to people all over the world.

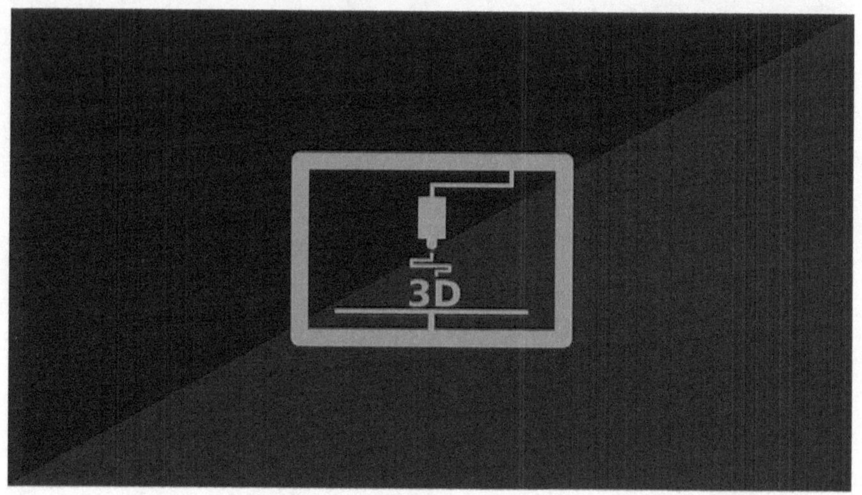


### 3D Printing

The invention of 3D printing is often attributed to Chuck Hull, who in 1986 developed a technology called stereolithography (SLA) that used a UV laser to solidify photopolymer resin layer by layer to create 3D objects. Hull went on to found 3D Systems, one of the first companies to commercialize 3D printing technology. However, it's worth noting that the development of 3D printing was a collaborative effort, and there were many other researchers and inventors who contributed to its development over the years. 3D Printing - Invented by Chuck Hull in 1986, but first commercially introduced in Japan in 1988.

## Karaoke

Karaoke was invented by a Japanese musician named Daisuke Inoue in the early 1970s. Inoue created a machine that played instrumental versions of popular songs and displayed lyrics on a screen, allowing people to sing along with the music. He initially rented out the machines to bars and clubs in Japan, and the concept quickly became popular throughout Asia and around the world. Today, karaoke is enjoyed by millions of people in countries all over the globe.

## The Suitcase

The concept of carrying personal belongings while traveling dates back to ancient times, but the modern suitcase as we know it today was invented in the late 19th century.

The first suitcases were made of leather and were heavy and cumbersome. In 1877, a man named Henry S. Hall patented the first "portable wardrobe," a flat-top trunk with a hinged lid that could be opened like a suitcase. The trunk was designed to hold clothes and was equipped with hangers and hooks for hanging garments.

In 1892, a man named Jesse Shwayder founded the Shwayder Trunk Manufacturing Company, which later became Samsonite Corporation. Shwayder introduced the first suitcase with a hard shell made of vulcanized fiber, which was lightweight, durable, and waterproof.

Over the years, suitcases have continued to evolve and improve, with new materials, designs, and features. Today, suitcases come in a variety of sizes and styles, from small carry-on bags to large checked luggage, and are an essential part of travel for millions of people around the world.